Maria Lucineide Gomes da Silva
Girlaine S.S. Alencar
Sóstenes G. Sousa

The unsustainability of vegetable production in Crato-CE

Maria Lucineide Gomes da Silva
Girlaine S.S. Alencar
Sóstenes G. Sousa

The unsustainability of vegetable production in Crato-CE

Marked by the abusive use of pesticides

ScienciaScripts

Imprint

Cover image: www.ingimage.com

This book is a translation from the original published under ISBN 978-3-330-76659-4.

Publisher:
Sciencia Scripts
is a trademark of
Dodo Books Indian Ocean Ltd. and OmniScriptum S.R.L publishing group

120 High Road, East Finchley, London, N2 9ED, United Kingdom
Str. Armeneasca 28/1, office 1, Chisinau MD-2012, Republic of Moldova, Europe
Managing Directors: Ieva Konstantinova, Victoria Ursu
info@omniscriptum.com

Printed at: see last page
ISBN: 978-620-8-53979-5

SUMMARY

Family farming is currently considered to supply both the domestic and foreign markets and is the largest producer of vegetables in the country. Brazil is a major producer of vegetables, with an annual turnover of more than R$10 billion. However, pesticides are used indiscriminately and these products are highly damaging to the environment and human health.

Ceará is the fourth Brazilian state with the highest number of establishments selling these products. The aim of this study was to assess the socio-economic profile of producers and identify the types of pesticides used on vegetable plantations in the municipality of Crato-CE. It was carried out in a descriptive manner, with technical expeditions to the production sites of coriander, lettuce, chilli peppers and spring onions. It was found that the production model used in horticulture in the municipality does not have major technological advances and the properties are fairly homogeneous, 84% of the workers are male, 67% of the properties use spring water to irrigate the crops. We observed the use of around five pesticides, none of which are recommended for the crops, such as cypermethrin, NORTOX 250 EC and AGRITOATO 400. The applicators don't use Personal Protective Equipment (PPE) and haven't been trained to do the job. There is little supervision and technical support for vegetable growers. Vegetable production in the municipality is fundamental for generating income for the agents involved, but it can also generate significant socio-environmental impacts such as contamination of the soil, water and air, and directly affect the health of the workers. It can also cause long-term problems for consumers, due to the probable contamination of fruit with pesticide residues. In this context, emergency measures are needed to minimise the use of pesticides in horticulture in the municipality of Crato, as well as the need for alternative means of controlling vegetable pests and diseases, so that producers are less exposed to poisons and can supply clean, quality products.

Keywords: Poisons, Public health, Healthy food.

Summary

CHAPTER 1

INTRODUCTION

The term horticulture comes from the Latin hortus (garden + culture), which means the art of cultivating gardens (ALENCAR, 2012).

Family farming, once known as "subsistence work" (emphasis added), is now considered to supply both the domestic and foreign markets, with significant figures in terms of the percentage that this production represents at national level (GRECZYSZN, 2013).

This sector is characterised by being responsible for generating temporary jobs and is the largest producer of vegetables in the country (MELO, 2012).

In Brazil, it is estimated that under normal market conditions, vegetables generate an income of between US$ 2,000 and US$ 20,000 per hectare cultivated in open fields. 55% to 60% of the volume of vegetables is sold by wholesalers, which generates an annual turnover of 15 million tonnes, totalling a wholesale value of more than R$10 billion (SEBRAE, 2015).

However, with growing demand and the requirement for better quality products, vegetable production in our country has shown distinct characteristics between regions, revealing major differences in relation to the adoption of inputs and technologies aimed at improving the sector (CRUZ, 2010).

The figures for pesticide consumption in the country are alarming. In recent years, consumption has increased by 700% (EMBRAPA, 2015). The pesticide industry in the country sold 708,000 tonnes of inputs in 2010, 730,000 tonnes in 2011 and 823,000 tonnes in 2012 (PORTELA, 2016).

Pesticides are compounds which, due to the substances present in their composition, acquire a certain degree of toxicity, ranging from slightly to extremely toxic, which makes them highly harmful to the environment. These products are used to eliminate pests and crop diseases. However, when they are used indiscriminately, they can directly contaminate humans and the workers who handle them, as well as contaminating the soil, air, groundwater and surface water, causing serious damage to the environment. It can also reach the table of consumers of vegetables and fruit in general, due to the risk of leaving pesticide residues in food.

According to Colares (2011), Ceará is the fourth largest Brazilian state in terms of the number of establishments that sell pesticides. This is due to the tax exemption for pesticides within the state, highlighting the existence of legislation that reduces these taxes to 60%. This "benefit" could be extended to production companies.

Studies carried out by Silva et al. (2011) in the Cariri Metropolitan region found that 80% of the pesticides used on farms were not indicated for horticulture and 20% of the products were used at dosages above those recommended by ANVISA. When pesticides are used that are not recommended for the crop, as well as being illegal, the food will be more easily contaminated and may have pesticide residues, losing quality and safety in its production (ANVISA, 2011).

Pesticide residues in food can have short- and long-term effects on consumers, since there is a lack of conditions for detecting signs of pesticide poisoning and false-negative diagnoses are produced, preventing early diagnosis and the subsequent start of treatment. In Brazil, several states have already recorded episodes of direct or indirect poisoning with deaths, thus constituting a serious public health problem (CARNEIRO et al., 2011).

In this context, there is a need for local studies to assess socio-environmental problems in vegetable production in the municipality of Crato-CE.

CHAPTER 2

LITERATURE REVIEW

1.1 The use of pesticides and their effects on human health

During the 1970s, there was a "leap" in the use of pesticides in Brazil, which was driven by agricultural financing, which included a package of poisons, fertilisers and chemical fertilisers (ARAÚJO et al., 2007; PALMA, 2011). This led to the mass dissemination of these products among producers, using the pretext of increasing production and productivity.

This fact is highlighted by Mata and Ferreira (2013), who point out that from 1970 onwards, national agricultural production had to adapt to this new policy of massifying the use of pesticides, due to the compulsory successive applications of these products on crops that had bank financing, on the grounds that this methodology would safeguard the productive potential of the crops.

These practices have had negative social and environmental consequences, including water and soil pollution, contamination of workers and the loss of biodiversity due to the indiscriminate use of these chemical products (ALENCAR, 2006).

Various studies have shown that the use of pesticides in Brazil is often indiscriminate and harmful to the health of the population (CRUZ, 2014). The inappropriate use of these products stems from a number of factors, including inefficient inspection (WAICHMAN, 2012), which favours the excessive use of authorised products or the application of active ingredients that are not permitted for use in the country (BRASIL, 2010, 2011, 2013), the poor technical guidance available for plant and animal production (FARIA; FASSA; FACCHINI, 2007; LONDON, 2011) and lenient regulations at regarding the approval of the use of products that have already been banned in other countries due to the potential risk to human health (PELAEZ; SILVA; ARAÚJO, 2012).

Peres et al. (2005) add that in rural Brazil, the inappropriate use of pesticides has caused serious and complex health problems for workers, ranging from simple headaches and dizziness to fainting and convulsions. Another fact addressed by the authors is that the means of contamination is not only through direct contact with the products, but also through contact with the environment or contaminated objects and contamination of the biota in areas close to agricultural plantations. However, they emphasise that the processes through which human populations are exposed to the products are a real mystery, given the multiplicity of factors involved, which is why they often do not attribute the problem to the use of pesticides.

Pesticide residues left on food is another major problem resulting from the abusive use of these products. According to Pedroza (2013), a study carried out by the Programme for the Analysis of Pesticide Residues in Food (PARA) in 2010 found that samples collected in 26 Brazilian states showed that a third of the food consumed daily by Brazilians is contaminated by pesticides. Of the samples analysed, 63% were contaminated with pesticides and 35% had active ingredients that were not authorised (NA) for that crop and/or exceeded the maximum residue limits (MRL).

According to ANVISA (2010), among the ingredients found are Indosulfan, Metamiclofós and Acephate, which have a high degree of proven acute toxicity and cause neurological and reproductive problems, hormonal disruption and even cancer. These are banned in many parts of the world, such as the European Union and the United States, yet they continue to be commercialised in Brazil.

The use of agrochemicals is not only related to agricultural production, as we can see in the work carried out by Alencar (2006; 2013) with flower growers in the Cariri region of Ceará, who reported that the main problems in relation to agrochemicals are: lack of technical information, inadequate use or non-use of personal protective equipment (PPE) when applying the products, inadequate dosages (above or below those recommended), use of poisons not indicated for the crops and inadequate disposal of empty containers.

There has been an exponential growth in contamination caused by improper handling of pesticides in the field, mainly due to a lack of knowledge on the part of farmers. In Brazil, strategies to control and prevent pesticide poisoning are inexpressive and inefficient, due to the lack of specialised professionals in this area (MOURA, 2005). As a result, many cases go unnoticed by the health authorities.

The search for control and elimination of unwanted species in crops has led farmers to start using pesticides indiscriminately. This has contaminated the entire environment, threatening all species (GEREMIA, 2011).

The big problem is that most farmers, especially those who don't have technical knowledge, consider pesticides to be a remedy, something beneficial that will increase their productivity, leaving aside all the negative consequences for the environment and their own health, most of which they don't know about. The best and most efficient way to change this agricultural system rooted in the use of "poisons" ***is to*** change the customs that have been implanted throughout society, combined with good environmental education (PERES; MOREIRA; DUBOIS, 2003; BEDOR, 2008).

It should be emphasised that the scale of the impacts can gradually increase and become even worse when these inputs are used intensively in an uncontrolled way, because they are not selective and, as well as eradicating pests from the target environment, they also eliminate their natural enemies, i.e. their predators and competitors (SOARES, 2010).

The way in which pesticides act on "pests" is due to the presence in their composition of a toxic chemical molecule which is capable of inhibiting the normal biological activity of beings sensitive to it (TERRA, 2008), demonstrating its lethal capacity for the environment it comes into contact with.

It should be emphasised that the effects of pesticides go beyond what is commonly exposed, so it can be said that the effects on health do not only concern exposed workers who have direct contact with the products, but also the general population and the environment (ROSA, 2008).

Due to their persistence in the environment and the great potential to produce adverse health effects, pesticide contamination of water resources affects both surface and groundwater, which has been recognised as an important issue in many countries because it is a common good for all, since water scarcity caused by natural and anthropogenic processes is already a serious problem in many parts of the world. Due to their persistent, mobile and toxic characteristics, pesticides tend to accumulate in soil and biota. Pesticide residues can reach surface water through run-off and groundwater through leaching. When we analyse this interaction, we see that residues can pass through all levels of the food chain until they reach human beings (GORZA, 2012).

When we analyse the diseases caused by these poisons today, we see that they are increasingly frequent and silent due to the great potential for accumulation. The effects on human health can be acute or chronic. Acute effects result from exposure to concentrations of one or more agents that can cause almost immediate damage within twenty-four hours of exposure. This takes the form of allergies, muscle spasms, nausea, fainting, vomiting, convulsions, alterations to the nervous system and damage to the respiratory system. Chronic effects can cause long-term infertility, impotence, abortion, congenital malformations, hormonal dysregulation, effects on the immune system, cancer, among others (BOCCOLINI, 2010).

Despite studies showing the harmful effects of pesticides, their use is frequent in horticulture and agriculture worldwide. In Brazil, this problem is even more serious, as inspections are inefficient and the poisons are called "pesticides" by farmers. The country is also the world's leading consumer of pesticides, followed by the United States, China, Japan, France and Germany.

The sale of these poisons in the five countries together totalled US$28.2 billion (MCDOUGALL, 2013). Between 2013 and 2014, around 1 billion litres were used in Brazil, generating an average of 5 litres of pesticides per inhabitant. This intensifies the problems caused indirectly by the consumption of food without knowing where it came from (ROSI, 2015).

The risk of consuming food contaminated by pesticides creates serious problems for human health, many in the long term (ANVISA, 2006).

Jardim et al. (2009) emphasise that this contamination tends to be greater in foods consumed "in natura" and in their shell, as all pesticide residues involve risks that can be cumulative and even unknown to health.

According to the National Health Surveillance Agency (ANVISA), 25% of the food samples analysed in the country showed unsatisfactory results due to the presence of unauthorised products and when authorised, they are in concentrations above the permitted limit (PARA, 2014).

1.2 Vegetable production in Brazil

Vegetable production in the country has increased by 33% in the last ten years, while the area used to grow this type of food has decreased by 5% and productivity has increased by 38%. Around three quarters of the production volume is concentrated in the South East and South regions, while the North East and Centre West account for the remaining 25%. In the northern states, consumer markets are supplied by products from the north-east and south-east (MELO, 2007).

According to data from the IBGE (2005), total vegetable production in 2005 was 17,385.9 thousand tonnes, covering an area equivalent to 785.2 thousand hectares. The total value of production was estimated at R$11,482.42 million. It should be noted that only six vegetables account for more than 64% of the total volume produced: tomatoes, potatoes, watermelons, onions and sweet potatoes (Table 01).

Vegetables	Production (thousand tonnes)	Share (%)
Tomatoes	3.303,5	19,0
Potato	2.951,0	17,0
Watermelon	1.850,0	10,6
Chives	1.059,0	6,0
Carrot	765,8	4,4
Sweet potato	538,5	3,1
TOTAL	17.385,9	100,0

Table 01: Vegetables produced in Brazil in 2005. Source: IBGE, 2005.

Another peculiar characteristic of this sector highlighted by MELO (2007) is that the vast majority of vegetables produced (60%) are concentrated on family farms of less than 10 hectares. As an agro-economic activity, it also stands out for requiring high levels of investment, unlike other agricultural activities. However, this activity makes it possible to obtain high yields per hectare cultivated and per hectare/year, depending on the added value of the product and the market situation.

The growth of this market in the country is evident: in 2011 the area under cultivation was 946,000 hectares, with an estimated production volume of around 19.4 million tonnes. It is believed that the sector was responsible for generating 2.4 million direct jobs, with the estimated value of the labour employed being 1.6 billion reais (ABCSEM, 2011).

CHAPTER 3

STUDY AREA

The municipality of Crato is part of the Cariri Metropolitan Region located in the south of the state of Ceará, created by Complementary Law 78 of 29 June 2009, which brings together the following municipalities: Crato, Juazeiro do Norte, Barbalha, Nova Olinda, Santana do Cariri Farias Brito, Caririaçu, Missão Velha and Jardim, as can be seen on Map 01:

MAPA 01- Região Metropolitana do Cariri

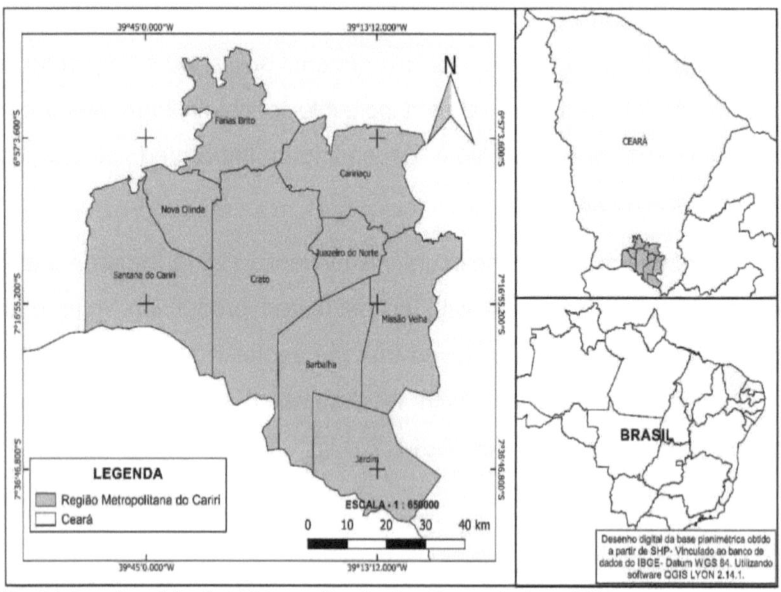

Map 01: Location of the Cariri Metropolitan Region. Source: Authors, 2016.

According to Queiroz (2013), the formation of the Cariri Metropolitan Region represents a demand from the local community, which is initially linked to the geoenvironmental characteristics that differentiate the fertile, spring-rich and prosperous valley from the rest of the arid northeastern hinterland.

This region stands out for its high potential for economic development, especially the municipalities that make up the CRAJUBAR Triangle (Crato, Juazeiro do Norte and Barbalha). These are the urban centres in the south of Ceará that concentrate most of the population and have the best regional socio-economic indicators (NASCIMENTO, 2015).

CRAJUBAR is formed by the conurbation of the municipalities of Crato, Juazeiro do Norte and Barbalha. It belongs to the Metropolitan Region of Cariri, which is located in the Cariri micro-region of Ceará and is an atypical geographical space in relation to the characteristics of the Brazilian semi-arid region. This micro-region is commonly referred to as an "oasis" in the sertão, due to its milder climate and its great availability of groundwater, as it is surrounded by the Bci do Araripe, where the Arripe National Forest is located.

The municipality of Crto, within CRAJUBAR, is the oldest, created in 1764. The presence of the Chpd do Arripe and the Arripe National Forest (FLONA) give the town its unique beauty and climate. The FLONA, created in 1946, is made up of "[...] the Serra Úmida/Cerrado ecosystem, whose characteristics include the presence of exuberant vegetation, microclimate formation, a large presence of natural springs and rich palaeontological sites" (PLANO DE GESTÃO SÓCIO-AMBIENTAL DO CARIRI CENTRAL - PROJETO CIDADES DO CEARÁ, 2008, p. 31). The FLONA is located on the edge of the Bci do Arripe, which is made up of the Chapada do Arripe, surrounding municipalities in Cerá and Pernmbuco.

Crto, which is formed on the slopes of the Chpd and has natural springs, stands out for its serene clubs, its potential for developing ecological and scientific tourism, as well as its cultural festivals, the most important of which is EXPOCRATO. The municipality in question was formed as a result of the agricultural exploitation of the fertile area of the Cariri Valley, and was one of the main trading posts in Ceará in the 19th century.

Located 507 km south of the capital, Fortaleza, the municipality has edaphoclimatic conditions (light, water and soil) suitable for growing vegetables, as well as being close to a major consumer centre, with approximately 540,000 people (IBGE, 2010 apud ALENCAR, 2012). All of these characteristics drive the development of agricultural activities such as the cultivation of beans and vegetables in general (SANTOS, 2013).

In this context, the aim of this study was to diagnose the horticultural sector in this city, analysing the socio-economic profile of farmers who grow vegetables in the municipality of Crato - CE, as well as their production methods, with an emphasis on the use of pesticides.

CHAPTER 4

MATERIAL AND METHODS

A survey of the municipality's vegetable producers was carried out with the extension agency Empresa de Assistência Técnica e Extensão do Ceará - EMATERCE (Ceará Technical Assistance and Extension Company) and at the town's municipal market, where some of the producers sell their vegetables at the open market.

Subsequently, data on the rural population and vegetable production in the municipality was collected from the Ceará Institute for Research and Economic Strategy (IPECE), the Brazilian Institute of Geography and Statistics (IBGE), among others. After these preliminary studies, expeditions were made to the identified sites in July 2015, making it possible to survey the relevant aspects, identified through photographic records. In order to draw up maps during these visits, the points were surveyed using a Garmim GPSMAP portable GPS, accurate to three metres, from the ecological studies laboratory at the Federal Institute of Ceará.

Subsequently, Google Earth Pro and Qgis Lyon 2.12.1 software were used to compile thematic maps showing the location of the properties visited (Map 02).

MAPA 02 - LOCALIZAÇÃO DOS SÍTIOS ESTUDADOS

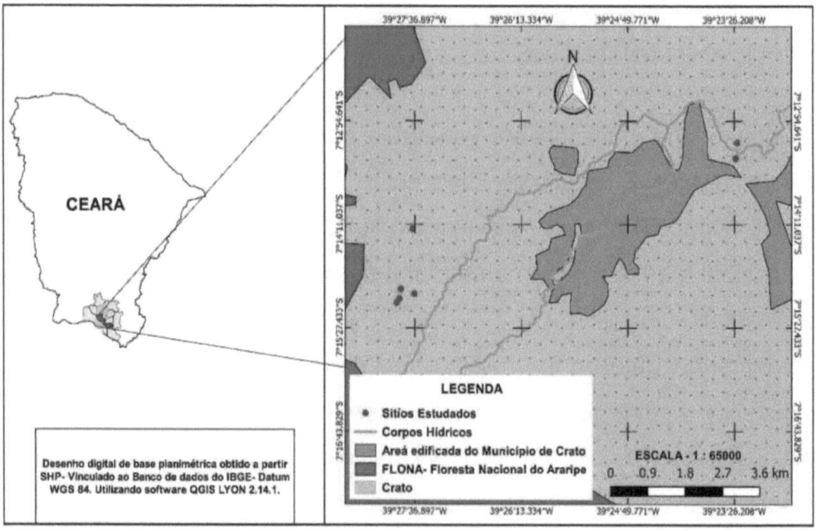

Map 02: Location of the sites studied. Source: Authors, 2016.

Many vegetable production sites were found, but some employees and owners did not want to take part in the survey and others were not on site, so we interviewed eight producers.

During the expeditions, semi-structured interviews were carried out with the agents involved, based on Alencar (2006) (Appendix A). The semi-structured interviews covered aspects such as: the characteristics of the farms, the socio-economic profile of the workers and producers, the consumer markets for the vegetables produced in the municipality, the health, safety and well-being of the workers, the environmental management carried out on the farms and a survey of the pesticides used on the farms.

After returning to the institute, the data was analysed by organising the data collected into tables and charts in order to make it easier to understand.

it was possible to obtain information for the socio-environmental analysis of vegetable producers in the municipality of Crato.

After surveying the pesticides, consultations were held with AGROFIT to check the technical recommendations and dosages for each vegetable grown on the farms.

Based on this information, it was possible to find out about the socio-environmental problems caused by growing vegetables in the municipality of Crato - CE.

CHAPTER 5

RESULTS AND DISCUSSION

5.1 Vegetables produced and the production model

The main vegetables grown in the municipality of Crato are: coriander, lettuce, chilli peppers and spring onions. Among the vegetables produced on the sites visited, coriander and spring onions stand out, both accounting for 29% of monthly production, in second place is lettuce (19%), followed by peppers and chilli peppers (Graph 01).

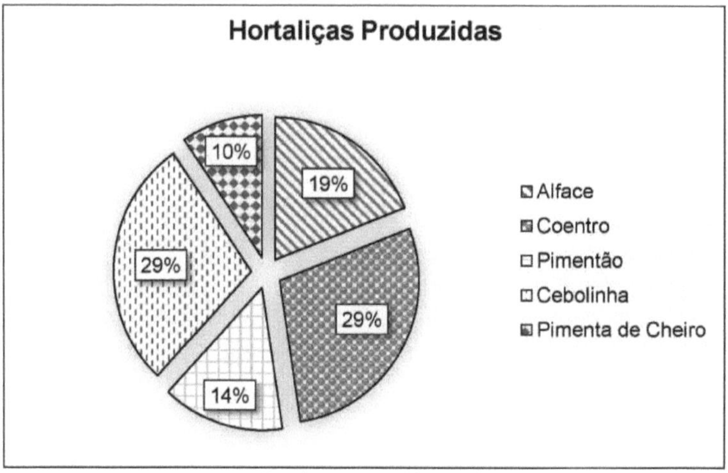

Graph 01: Vegetables produced. IFCE, 2015.

All these vegetables have a fast cycle, for example coriander has a cycle of 30 to 40 days, spring onions 60 days in summer and 90 in winter. The preference for these vegetables is due to their fast cycle, which directly influences the production and marketing of the vegetables. The production model used in horticulture in the municipality does not have major technological advances and the properties are fairly homogeneous. In most of the properties visited, there was a strong family relationship present in the organisation and throughout the production and marketing process of vegetables. There is usually a division of labour, where the men are in charge of building, cleaning and maintaining the area to be planted, while the women make the seedlings, transplant them, help maintain the beds and clean the vegetables for sale. When the vegetables are not sold at the open market, they are ordered by traders and small fruit markets in the city centre. However, the figure of the man, the father of the family, the owner of the house is present on all the properties. The data collected showed that all the owners are male, and among the employees, the number of men (84%) is higher than the number of women (Graph 02).

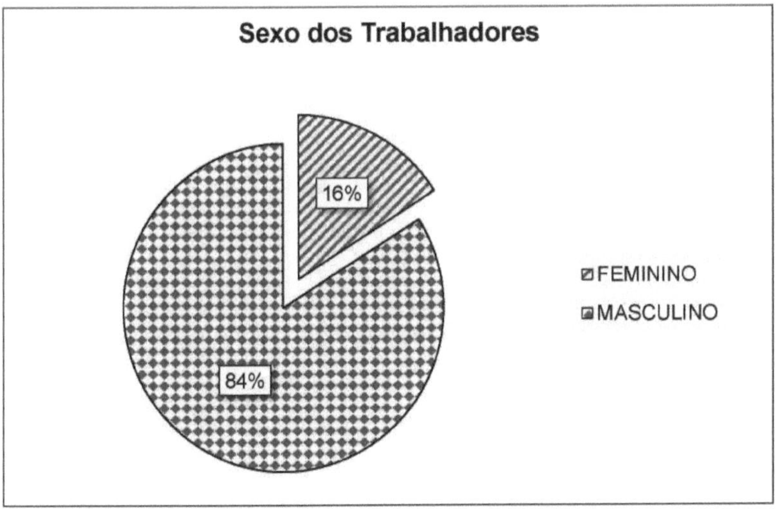

Graph 02: Sex of workers. IFCE, 2015.

With regard to the level of education of the agents involved, 38% are illiterate, but 48% have incomplete primary education and only 5% have secondary education (Graph 03). The technique used to build and maintain the properties is based on the farmers' tacit knowledge. This knowledge was acquired through the experience of the producers, or passed down from generation to generation, between parents and children. This explains the lack of major technological advances and the homogeneity of the production model adopted by the communities visited.

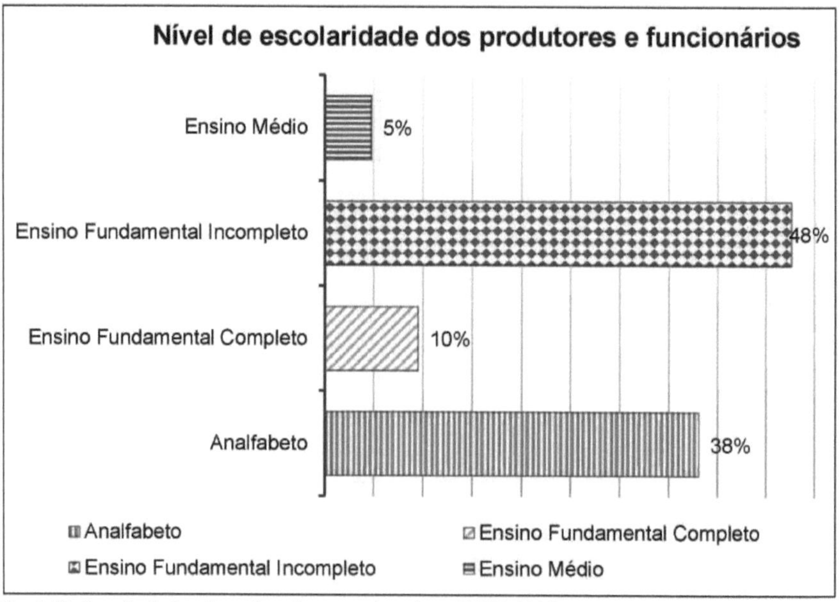

Graph 03: Level of education of producers and workers. IFCE, 2015.

The technique used for cultivation in the region is the open field (Figure 01), which is the system most practised by vegetable growers throughout Brazil. The factor driving the adoption of this type of cultivation is the soil and climate conditions present throughout the country, which allow crops to be grown all year round and thus meet the demand of consumer markets. However, this type of system is totally dependent on the climatic conditions of the environment in which it is located (BRANCO, 2014).

Figure 01: Open field cultivation. Source: Field research, 2015.

In 67% of the properties, spring water is used to irrigate crops (Graph 04) and none of them has a water meter or any other type of efficient control over water use. It's worth pointing out that on these properties there is no minimum control over the use of this resource, and the crops are irrigated much of the time, both during the day and at night.

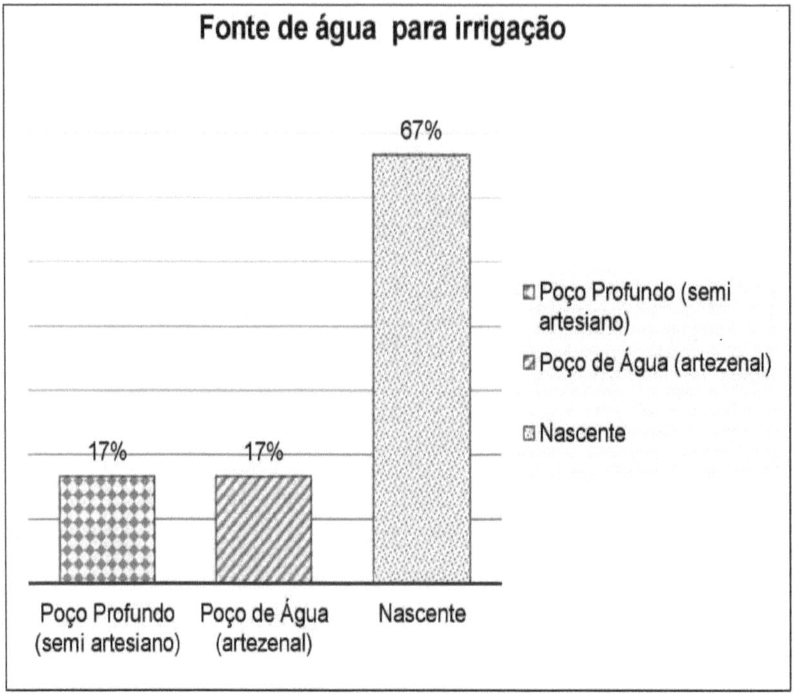

Graph 04: Source of water for irrigation. IFCE, 2015.

The technology most used for irrigation on 50 % of the properties is micro-sprinkling, followed by drip irrigation (33%), noting that some properties that use drip irrigation use hoses to help with irrigation (Graph 05).

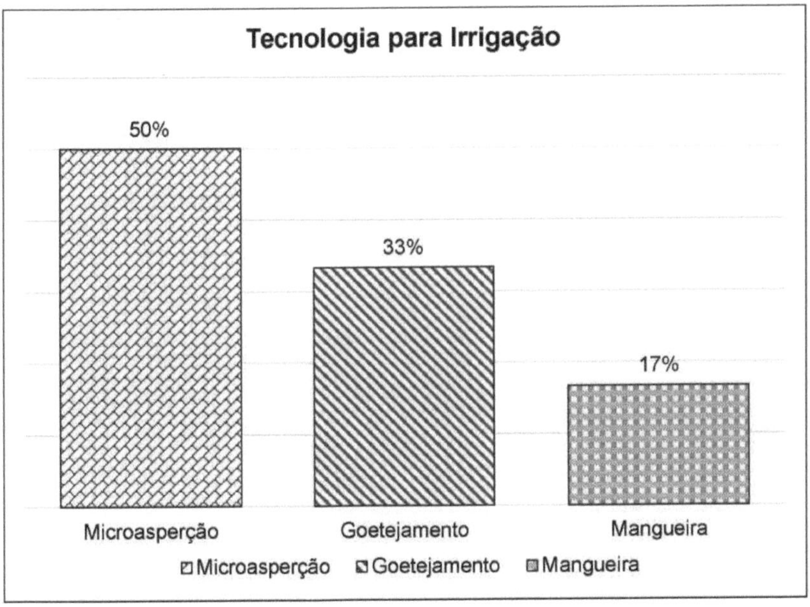

Graph 05: Technologies used for irrigation. IFCE, 2015.

The technologies used for irrigation on these properties are associated with the localised irrigation method, which saves water, reduces energy and labour, as well as boosting productivity and reducing vegetable diseases. However, on the properties that use micro-spraying, which simulates a small amount of rain, there is a great deal of wasted water, possibly due to a lack of guidance from the owners (Figure 02). It is believed that the drip system would be the most suitable for these producers, as it makes the water available close to the plant's root system, using a smaller volume of water, which reduces costs and increases productivity, as the plant receives the necessary volume of water to develop.

Figure 02: Irrigation model. Source: Field research, 2015.

It was found that they do not receive any kind of training or monitoring from a technical team that can provide specialised support to improve vegetable production in the municipality. None of the properties have adequate phytosanitary conditions, and there is a total lack of toilets, bathrooms and sinks for workers to sanitise, which intensifies contamination of the workers and the environment. With regard to the personal protective equipment (PPE) needed to handle and apply pesticides, only a few boots and gloves were found on one of the properties.

5.2 The pesticides used in vegetable production in the municipality

By analysing the data obtained, it was possible to observe the constant use of pesticides on the municipality's vegetable farms, which belong to various toxicity classifications, from IV (low toxicity) to I (extremely toxic) (Table 02).

Table 02. Pesticides used on farms. Database: AGROFIT,2016.

Agrotoxics	Chemical group / Active ingredient	MAP Registration No.	Class	Mode of application	Toxicological classification	Environmental Hazards
Cypermethrin NORTOX 250 EC	Cypermethrin - Pyrethroid	3101	Contact Insecticide	Land/ Air	1 - Extremely Toxic	II-Very Dangerous to the Environment
AGRITOATO 400	Organophosphate	00178810	Insecticide and acaricide	Land/ Air	1 - Extremely Toxic	II-Very Dangerous to the Environment
Roundup 0 Original	Substituted glycine	00898793	Herbicide	Land/ Air	III - Medium Toxic	111-Dangerous to the Environment
Mospilan	Neonicotinoid/ Acetamiprid	10498	Insecticide	Land/ Air	III - Medium Toxic	II-Very Dangerous to the Environment
Kumulus DF	Inorganic/ Sulphur	2418592	Acaricide and Fungicide	Land/ Air	IV - Low toxicity	IV - Not Very Dangerous to the Environment

The agrochemicals used on vegetables have various levels of environmental danger. The most toxic are the insecticides, Cypermethrin NORTOX 250 EC and AGRITOATO 400, the others are less toxic, being classified as very dangerous, dangerous and slightly dangerous.

It's worth noting that none of the agrochemicals used have a recommended use for the vegetables grown. AGRITOATO 400, for example, is an insecticide indicated for pest control in citrus crops, and in one of the properties visited this product is being applied directly to chilli and pepper crops. This product has a grace period of 3 days and requires an interval of 48 hours for people to re-enter the treated area. This grace period or safety interval is disregarded by some producers and unknown by others (NUFARM, 2010).

Cypermethrin nortox 250 EC is a pyrethroid insecticide that acts by contact and ingestion on biological targets that cause considerable damage to the production of cotton, rice, potatoes, coffee, citrus, beans, cassava, maize, soya and tomatoes and has a safety interval of between 10 and 30 days depending on the crop (NORTOX, 2010).

Roundup Original is used to eliminate weeds in cultivated areas (post-emergence of crops and weeds) in the following crops: plum, banana, cocoa, coffee, sugar cane, citrus, apple, nectarine, pear, peach, grape, pasture, pine and eucalyptus. This product has a safety interval of between 15 and 30 days depending on the crop, utility and application period (MONSANTO, 2009).

Kumulus DF is used against various species of mites and fungal development pathogens on citrus, apple, peach, bean, grape, courgette, pumpkin, cucumber, mango, papaya, soya and wheat crops. The safety interval for this pesticide ranges from 15 to 30 days for banana, cocoa, citrus, nectarine, peach (30 days); plum, grape (17 days) apple, coffee, pear (15 days); for rice, sugar cane, maize and wheat and for pine and eucalyptus because they are not used for food purposes (BASF, 2013).

The insecticide Mospilan, used by producers on lettuce plantations, is recommended for potato, tomato, cotton, wheat, melon, watermelon and bean crops, with a safety interval ranging from 3 to 15 days. This insecticide is highly easily infiltrated into the soil and can reach areas neighbouring the treated areas, groundwater and surface water, meaning that this product is highly persistent in the environment and has been applied indiscriminately in Crato's horticulture (IHARABRAS, 2011).

The use of pesticides on crops for which they are not authorised has negative consequences for human and environmental health, one of which is increased food insecurity for consumers who eat contaminated food (OLIVEIRA, 2014).

The irregularity in the way these agrochemicals are applied and prepared is alarming, carried out without any technical guidance. The pesticides are applied directly to the vegetables, without taking into account the instructions for use. No care is taken with the dosages for diluting the products, and in some cases they are applied by hand in small buckets. Producers believe that the greater the quantity of the product, the better the result.

According to OLIVEIRA (2005), regardless of the method of application, pesticides end up reaching the soil and water, whether through rainwater washing off the leaves, leaching or soil erosion. It should be emphasised that man is always the main recipient.

The damage to human health caused by the use of pesticides has constantly been the subject of various studies carried out by a wide range of professionals (SIQUEIRA, 2008).

Depending on the form, exposure time and type of product, pesticides can cause severe effects on human health. These effects can be acute, due to short-term exposure, or chronic, due to prolonged exposure. The main symptoms of acute effects are the rapid and clear signs of intoxication typical of the product or other adverse effects, such as skin lesions, irritation of the mucous membranes of the eyes, nose and throat, stomach pain (epigastralgia); in chronic effects, the effects are adverse and often irreversible (LONDON, 2011).

Cases of pesticide poisoning show an increase in suicide attempts, damage to the central nervous system, mutagenic, carcinogenic and teratogenic effects, as well as a series of other diseases and health complications (COSTA; ROHLFS, 2011).

With regard to storage, agrochemicals were found in plastic bags (FIGURE 03), going against the explicit instructions on the packaging of all agrochemicals, which make it clear that the product should always be kept in its original, unopened packaging.

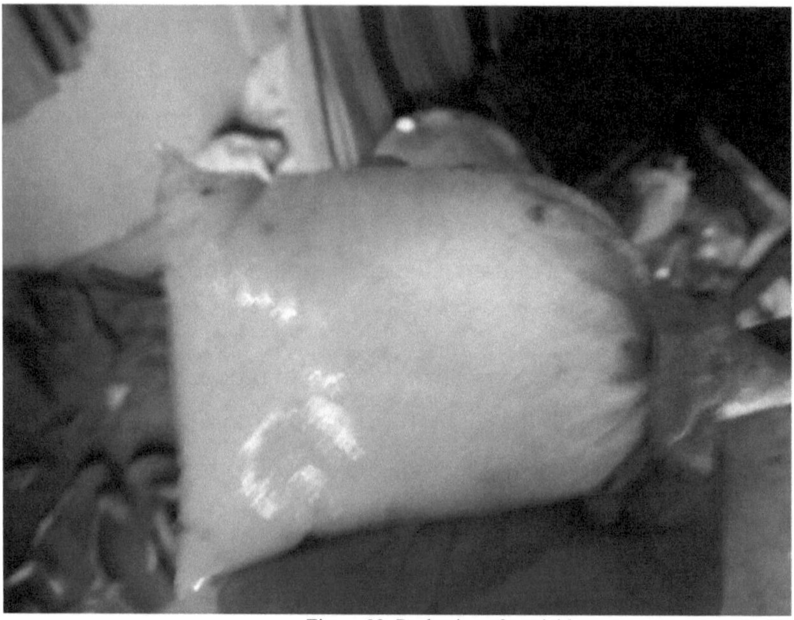

Figure 03: Packaging of pesticides
Source: Field research, 2015.

The room used to store these products in accordance with the technical recommendations must be made of masonry or non-combustible material, always ventilated, covered and impermeable, with warning signs saying: CAUTION VENO. This area must be locked to prevent unauthorised access, especially by children. The reality at the sites we visited was quite the opposite. They are usually packed in nylon bags and taken home at the end of the day's work, improvising tents made of cardboard, canvas and wood (FIGURE 04).

Figure 04: Place where pesticides are stored Source: Field research, 2015.

The pesticides are stored in "shacks" which also serve as a resting place for the workers, where they eat their meals and leave their clothes and work equipment (FIGURE 05). According to Law No. 7,802 of 11 July 1989, in compliance with current national standards, observations and instructions provided by manufacturers, as well as the safety conditions explained on the label and package leaflet, there must be an exclusive place for toxic products, isolated from food, drink, feed or other materials and with warning signs.

Figure 05: Interior of the pesticide storage area Source: Field research, 2015.

With regard to the disposal of empty packaging, it is generally left out in the open or buried near construction sites and water sources. These containers can contain product residues and contaminate the soil, water and air, harming fauna, flora and people's health.

It was noted that there is no reception centre for empty pesticide containers in the region and that everything producers need is provided by local businesses.

Many of the producers are unaware that they have to return the empty packaging to the establishment where the product was purchased, which is a mandatory action valid for up to one year from the date of purchase. Some of the producers triple-wash the packaging, but this procedure is carried out inside the plantation itself and next to the water source used to irrigate the crops, without any use of PPE. None of the properties have PPE such as boots, gloves, aprons, face shields or respirators. This equipment is indicated for use in both the preparation and application of the products in order to avoid direct exposure to the toxic products, preventing possible poisoning of these workers.

Of the properties visited, in four the workers claimed to have experienced headaches, dizziness, malaise, eye irritation and a burning sensation on the skin during and/or after the activity. On one of them, an employee reported having acquired ophthalmological problems after getting wet hands in the eye area from an agrochemical he was preparing without any kind of protection.

According to the National Cancer Institute (INCA, 2010), the most appropriate course of action would be not to use pesticides, as individual protection or local barriers are palliative control measures that do not prevent the spread and contamination of these substances over long areas. It also warns of their potential to cause cancer in the long term and intoxication in the short term. In addition to the known risks of poisoning, there are also other adverse effects resulting from the use of pesticides.

In recent years, substances including pesticides are suspected of interfering with the endocrine systems of both humans and wild animals and can cause harmful health effects such as cancer, behavioural changes and reproductive abnormalities (NODARI, 2012).

When analysing the chemical group and active ingredient of the two most toxic pesticides used (Cypermethrin NORTOX 250 EC and AGRITOATO 400), there is Cypermethrin - Pyrethroid, which causes adverse effects on the skin, such as irritation, numbness, tingling and can be easily absorbed by the human body via the skin, causing everything from dizziness to convulsions. An organophosphate is also found, considered by researchers to be a probable cause of neurobehavioural disorders. Studies in AGROSOFT (2005) point to at least 3 types of neurological sequelae after acute intoxication or due to chronic exposure: Delayed Polyneuropathy, Intermediate Syndrome and behavioural effects.

Organophosphate class pesticides are absorbed through the skin, respiratory tract and gastrointestinal tract, and their absorption is often favoured by the solvents present in the formulation. In exposures that occur during industrial manufacturing processes, formulation, agricultural application or vector control in public health, the main routes of exposure are respiratory and cutaneous. Cutaneous absorption is greater at high temperatures or when there are lesions on the skin. There are different absorption rates between the various types of compounds, but all can lead to poisoning if workers are not sufficiently protected. In cases of voluntary ingestion (suicide attempts) and involuntary ingestion (accidents, usually involving children, contamination of food and/or water, homicides), the main site of absorption is the digestive tract. Absorption by this route also occurs in workplaces with poor hygiene or when workers eat in contaminated areas (ALONZO, 2013).

Pyrethroids are synthetic insecticides made from natural pyrethrins and are mainly used as household products, in agriculture and in public health campaigns. Pyrethroid poisoning affects the nervous system, causing problems such as hypersensitivity to cutaneous sensory stimuli, tremors, hyperexcitability, excessive salivation (sialorrhoea), convulsions and paralysis.

These intoxications are related to their chemical structure. Synthetic insecticides can be structurally divided into two groups according to the absence or presence of a cyano group (CN) in the phenoxybenzyl portion, being (Type I) and (Type II) respectively. Type I pyrethroids act mainly on the peripheral nerves, causing Type I Poisoning Syndrome or the so-called "T Syndrome", known to cause, in rats, tremors all over the body, aggressive behaviour, increased sensitivity to external stimuli, hyperexcitability, ataxia and convulsions. In non-rodent mammals it causes progressive paralysis. Type II pyrethroids generally act on the central nervous system, producing type II choreoathetosis syndrome, or CS syndrome. In this case, the symptoms observed in rats were hypersensitivity, abundant salivation, shaking of the hands or forepaws, digging movements and periodic tremors that can evolve into choreoathetosis and, in some cases, repetitive clonic movements (SANTOS,2008).

CHAPTER 6

CONCLUSION

It was found that the procedures adopted to collect data and carry out the research provided knowledge of the production process, enabling an assessment of the socio-environmental problems encountered in growing vegetables in the municipality of Crato/CE.

Vegetable production is a very important activity for small farmers in the city of Crato-CE, as it is one of their main sources of income.

However, negative points were also highlighted, such as the improper use of agrochemicals in terms of dosage (high), indication for use and lack of training for applicators. Incorrect storage of agrochemicals and a lack of units for receiving empty containers. These points can generate significant socio-environmental impacts such as contamination of the soil, water and air and directly affect the health of workers and consumers. Consumers can suffer long-term problems due to the high level of contamination of vegetables with pesticide residues.

Several problems linked to the health and safety of plantation workers were identified, such as: lack of training for workers to carry out activities linked to the use and handling of pesticides with knowledge and safety, non-use of personal protective equipment (PPE) and inadequate storage of pesticides.

It was clear that producers and workers have little knowledge of the dangers that pesticides can cause to their health. The improper use of these products linked to the excessive use of water on the properties will lead to contamination of the environment, workers and the surrounding population.

The production system used in horticulture in the municipality has proved to be totally unsustainable from the point of view of environmental care. In all the properties analysed, the risks of environmental contamination are visible. Urgent measures are therefore needed to minimise the impacts generated.

There is a lack of technical support for producers. There is an urgent need for inspection of the municipality's horticultural crops and for continuous and efficient technical guidance aimed at proposing the replacement of pesticides with less toxic products, developing alternative means of containing vegetable pests and diseases.

It is hoped that this research will contribute to the sustainability of the vegetable production sector in the municipality of Crato/CE and that it will be a source of information to subsidise other research in this area. Among the recommendations for future work, it is suggested that this methodology be applied in other vegetable producing regions to identify the socio-environmental impacts generated and raise awareness among the agents involved.

For the study region, we suggest further studies aimed at analysing in detail the contamination of these producers by the use of pesticides, as well as an analysis of the amount of pesticide residues found in the vegetables produced and marketed in the municipality.

CHAPTER 7

REFERENCES

AGROFIT - Plant Protection Agrochemicals System. Information on the registration of pesticides and related products. Available at: <http://agrofit.agricultura.gov.br/agrofit_cons/principal_agrofit_cons>. Accessed on 22 Feb 2016.

ABCSEM - Brazilian Seed and Seedling Trade Association. Overview of the vegetable production chain in Brazil. Available at:< file:///C:/Users/lucin 000/Downloads/palestra10cursosementeshortalicas11%20(1).pdf>. Accessed on 22 July 2016.

ARAÚJO, F. V. Pesticide poisoning in rural workers: a literature review. 2010. 67 f. Monograph (Specialisation in Nursing). State University of Ceará. Fortaleza, 2010.

AGROSOFT. Pesticides: main symptoms of poisoning and diagnosis. General Editor, 29 June 2005. Available at :< http://www.agrosoft.org.br/br/agrotoxicos-principais-sintomas-de- intoxication-and-diagnosis/technology>. Accessed on 28 July 2016.

ALENCAR, G. S. S. Flower production and the EUREPGAP certificate: the case of Agropolo Cariri - Ceará. 2006. 153 f. Dissertation (Master's in Production Engineering) - Federal University of Paraíba. João Pessoa, 2006.

ALENCAR, G.S.S. Code of conduct: a potential for the sustainable development of floriculture in Agropolo Cariri/CE. 2013. 161 f. Thesis (Doctorate in Geography) - Paulista State University. Rio Claro - SP, 2013.

ALENCAR, S. K.P.; ALENCAR, G.S.S.; SANTANA, F.V.; ALENCAR.F.H.H.; SANTOS, C.A.A.; JUNIOR, E.R Diagnosis of Horticulture in the Metropolitan Region of Cariri/CE. In: VII Connepi-North-East Congress of Research and Innovation, 2012.

ALONZO, H. G. A.; CORRÊA, C.L. Praguicidas p.5,16 - Available at <http://www.saude.rs.gov.br/upload/1417454923_Cap%20Praguicidas%20Fundamentos%20Toxicologia%202014%20.pdf> Accessed on: 07 Sep. 2016.

ANIVA. Pesticide residues in food. Technical Report. Revista Saúde Pública, Brasília, v.40, n.2, p. 361-363, 2006. Available at: <http://www.scielo .br/pdf/rsp/v40n2/28547.pdf>. Accessed on: 16 July 2015

ANVISA. Primer on pesticides. Field trail series. 19 September 2011. Available at:<http://portal.anvisa.gov.br/documents/111215/451956/Cartilha+about+Agrot%C3%B3xicos+S %C3%A9rie+Trilhas+do+Campo/6304f09d-871f-467b-9c4a-73040c716676>. Accessed on 29 July

2016.

ANVISA. Programme for the Analysis of Pesticide Residues in Food. Brasília, 05 Dec. 2011. 2010 Activities Report. Available at: <http://portal.anvisa. gov.br/wps/wcm/connect/b380fe004965d38ab6abf74ed75 8 91ae/Relat%C3%B3 rio+PARA+2010+-+Vers%C3%A3o+Final.pdf?MOD=AJP ERES>. Accessed on: 20 March 2015.

ANVISA. Agrotoxics Report. 2012 Report. Available at: < http://portal.anvisagov.br/wps/wcm/connect/d6710700463436858 3a5bfec1b28f937/Relat%C3%B3rio+PARA+2012+2%C2%AA+Etapa+-+17_10_14-Final.pdf?MOD=AJPERES >Accessed on 13 July 2015.

BASF BRASIL. Bula: Kumulus DF. São Paulo, Brazil.2013.

BRAZIL, Law No. 7802, of 11 July 1989, Agrotoxics.

BEDOR, C. N. G. Study of the carcinogenic potential of pesticides used in fruit growing and its implications for health surveillance. 2008. 115f. Dissertation (Doctorate in Public Health) - Oswaldo Cruz Foundation. Recife, 2008. Available at: <http://157.86.8.70:8080/ certifica/bitst ream /icict/45465/2/97 1.pdf>. Accessed on: 20 July 2015.

BOCCOLINI, P. M. M. Exposure to pesticides, agricultural activity and mortality from non-Hodgkin's lymphoma in Brazil. 2010. 70 f. Dissertation (Master of Science in Public Health and Environment) - Oswaldo Cruz Foundation. Rio de Janeiro, 2010. Available at: <http ://www.saudepublica.ufc.br/ima gens/u ploads/dissertacoes/172def51d4ce042df50c8bfd178f2ba6.pdf>. Accessed on: 23 July 2015.

CARNEIRO, F.F., ALMEIDA, V.E.S, TEIXEIRA, M.M., BRAGA, L.Q.V. Agronegócio e Agroecologia: desafios para a formululação de políticas públicas sustentáveis In:RIGOTTO, R (Org) Agrotóxicos, Trabalho e Saúde - vulnerabilidade e resistência no contexto da modernização agrícola no Baixo Jaguaribe/CE. Fortaleza: Federal University of Ceará, 2011, p. 1-612.

CORRÊA, C. M. D. Effect of soya oil on the persistence of endosulfan in the environment. Piracicaba, 2005, p. 2005. Thesis (Doctorate in Agroecosystem Ecology) - Luiz de Queiroz College of Agriculture, University of São Paulo, Piracicaba, 2005.

COLARES, G. Brazil is the nation that consumes the most pesticides. Environment magazine. Available at: < http://www.revistameioambiente.com.br/2011/08/23/brasil-e-a-nacao-que-mais- consome-agrotoxicos/>. Accessed on 22 December 2015.

COSTA, F.L.F.; ROHLFS, D.B. Pesticide residues in food - implications for health and the environment. Pontifical Catholic University of Goiás. Postgraduate Programme in Forensic Bioscience. 2011.

CRUZ, G. The contamination of fruit, vegetables and greens (FLVS) with pesticide residues in Brazil and emerging opportunities. Postgraduate Programme in Agribusiness at Universidade Federal do Rio

Grande do Sul. Porto Alegre, 2014.

Available at: <http://www.agencia.cnptia.embrapa.br/gestor/agricultura_e_meio_ambi ente/arvore/CONTAG01_40_210200792814.html>. Accessed on 12 July 2016.

FARIA, N. M. X.F.; FASSA, A. G.; FACCHINI, L.A. Pesticide poisoning in Brazil: official information systems and challenges for epidemiological studies. Ciência & Saúde Coletiva, Rio de Janeiro, v.12, n.1 2007.

GEREMIA, B. Agrotoxics: The indiscriminate use of chemical products in the rural work environment and liability for damage to health. 2011. 147 f. Dissertation (Master of Laws) - University of Caxias do Sul. Caxias do Sul, 2011. Available at: <http://tede. ucs.br/tde_busca/arquivo.php?codArquivo =471>. Accessed on: 20 July 2015.

GORZA, N. L. Removal of pesticides in a conventional water treatment pilot plant associated with pre-oxidation and adsorption on granular activated carbon. 2012. 133 f. Dissertation (Master's in Environmental Engineering) - Federal University of Espírito Santo. Vitoria, 2012. Available at: <http://porta is4.ufes.br/posgrad/teses/tese_6155_ disse rta%E7% E3o%20N adja%2 0Lima%20Gorza.pdf>. Accessed on: 20 July 2015.

IBGE. Municipal Agricultural Production (PAM), 2005, Rio de Janeiro, IBGE. Available

www.sidra.ibge.gov.br, accessed on 20/11/2006.

IHARABRAS S/A INDÚSTRIAS QUÍMICAS. Leaflet: Mospilan. São Paulo, Brazil. 2011.

NATIONAL CANCER INSTITUTE. Coordination of Prevention and Surveillance. Surveillance of work-related and environmental cancer. 2. ed. rev. atual. Rio de Janeiro: INCA, 2010.

JARDIM, I. C. S. F. et al. Pesticide residues in food: a global environmental concern - a focus on apples. Química Nova, v. 32, n. 4, p. 996-1012, 2009. Available at: <http://www.scielo.br/pdf/qn/v32n4/v32n4a31.pdf>. Accessed on: 16 July 2015

LONDRES, F. Agrotóxicos no Brasil. Rio de Janeiro: AS-PTA, 2011.

MATA, J.S.; FERREIRA, R.L. Agrotoxics in Brazil - use and impacts on the environment and public health. EcoDebate: Citizenship & the Environment. Available at: <http://www.ecodebate.com.br /2013/08/02/agrotoxico-no-brasil-uso-e-impactosao-meioambiente- e-a-saude-publica-por-joao-siqueira-da-mata-e-rafael-lopesferreira/>. Accessed on: 07 Feb. 2016.

MCDOUGALL. P. The 10 countries that use the most pesticides in the world. Available at: <https:// lista10.org/diversos/os-10-paises-which-utilise-agrotoxicos-no-mundo/>. Accessed on: 24 July 2016

MONSANTO DO BRASIL LTDA. Bula: Roundup Original. São Paulo, Brazil.2009.

MELO, P.C.T.; VILELA, N.J. Importance of the Brazilian vegetable production chain. 2007.

MOURA, N. N. Perceived risk of pesticide use: the case of producers in São José de Ubá/RJ. 2005. 100 f. Dissertation (Master's in Development, Agriculture and Society) - Federal Rural University of Rio de Janeiro. Rio de Janeiro, 2005. Available at: <http:// r1.ufrr j.br/cpda /wpcontent/ uploads/2011/09/m _nara_nubia_d e_moura _2005.pdf>. Accessed on: 23 July 2015.

NASCIMENTO, D.C. Metropolitan Region of Cariri - CE: A Scenario of Uncertainties Regarding Management, Planning and Purposes. XI- ANPEGE National Meeting. ISSN 2175-8875. 2015.

NODARI, Rubens Onofre. Risk to the health of living beings from pesticides - emphasis on herbicides. In: ARAGAO, Alexandra. LEITE, José Rubens Morato; FERREIRA, Jovino dos Santos Ferreira; FERREIRA, Maria Leonor Paes Cavalcanti. Pesticides: our health and the environment at stake - technical, legal and ethical aspects. Florianópolis: Funjab, 2012.

NORTOX S/A. Leaflet. Cypermethrin nortox 250 EC. São Paulo, Brazil.2010.

NASCIMENTO, D,C. cariri metropolitan region - ce: a scenario of uncertainties regarding management, planning and purposes. In: XI-Anpege National Meeting, 9 to 12, 2015, Presidente Prudente, SP. Proceedings... Presidente Prudente, 2015.

NUFARM INDÚSTRIA QUÍMICA E FARMACÊUTICA S/A. Package leaflet: Agritoato 400. Ceará, Brazil.2010.

OLIVEIRA, S. de S. O papel da avaliação de riscos no gerenciamento de produtos agrotóxicos: diretrizes para a formululação de políticas públicas. São Paulo, 2005. 236 p. Thesis (Doctorate in Public Health - Postgraduate Programme in Public Health) School of Public Health - University of São Paulo (USP).

OLIVEIRA, L, C,C.Pesticide residues in food, a public health problem. Course Conclusion Paper presented to the Specialisation Course in Primary Care in Family Health, Federal University of the Triângulo Mineiro. Uberaba, Minas Gerais, 2014.

PALMA, D. C. A. Pesticides in human milk from mothers living in Lucas do Rio Verde - MT. 2011. 104 f. Dissertation (Master's in Collective Health) - Federal University of Mato Grosso. Cuiabá, 2011. Available at: <http://www.u fmt. br/ppgsc/ arquivos/857ae0a5ab2be9135cd279c 8ad4d4e61.pdf>. Accessed on: 23 July 2015.

PARA. Complementary Report on the Second Stage of Analyses of Samples Collected in 2012.Brasília. Anvisa 2014.

PELAEZ, V.S; SILVA, L.R., L. ARAÚJO, E. AGROTOXIC REGULATION: a comparative analysis. [S.I], SBHC, [2012].

PEDROZA, E. H. O. The indiscriminate use of pesticides and the violation of the fundamental rights to healthy food, health and a balanced environment and their consequences for social security. Justiça do direito. v. 27, n. 1, jan./jun. 2013. p. 220-233. Available at: <www.u pf.br/seer/index.php/rjd/article/

download/4566/3135>. Accessed on: 07 Feb. 2016.

PERES, Frederico; OLIVEIRA-SILVA, Jefferson José; DELLA-ROSA, Henrique Vicente; LUCCA, Sérgio Roberto de. Challenges to the study of human and environmental contamination by pesticides. Ciênc. saúde coletiva [online]. 2005, vol.10, p. 27-37 . Available at: <http://www.scielo.br /scielo .php?s cri pt=sci_arttex t&pid=S1413- 81232 0050 00500006&lng=en&nrm=iso>. Accessed on: 20 July 2015

PERES, F.; MOREIRA, J. C. É veneno ou é remédio? agrotóxicos, saúde e ambiente. Rio de Janeiro: Editora FIOCRUZ, 2003. 384 p. Available at: <http:/ /static.scielo.org/scielobooks/sg3mt/pdf/peres-9788575413173.pdf>. Accessed on: 14 Apr. 2015

CARIRI CENTRAL SOCIO-ENVIRONMENTAL MANAGEMENT PLAN. Ceará Cities Project. Fortaleza: Regional Hubs Urban Development Programme, 2008.

PORTELA, G.; TOURINHO, R. The power of legal and illegal pesticides in Brazil. Available at :< https://portal.fiocruz.br/pt-br/content/forca-dos-agrotoxicos-legais-e-ilegais-no-brasil>. Accessed on 26 Feb. 2016.

QUEIROZ, I. S. A metrópole fora do eixo: a emergência da região metropolitana do Cariri cearense no âmbito estadual. XIII National Symposium on Urban Geography, UERJ. In: Proceedings... Rio de Janeiro: 2013. Available at: <http://www.simpurb2013.com.br/wpcontent/uploads/2013/11/1652_Ivan.pdf>. Accessed on: 17 September 2014.

ROSI, M. The "alarming" use of pesticides in Brazil affects 70 per cent of food. El País. São Paulo. Apr. 2015. Available at: <http://brasil. elpais.com/brasil/2015/04/ 29/politica/1430321822 _851653. html>. Accessed on: 23 July 2016.

ROSA, I. F. Our meter is us, who feel and shout: socio-environmental conflict around an agrotoxics factory in Ceará. 2008. 205 f. Dissertation (Master's in Public Health) - Federal University of Ceará. Fortaleza, 2008. Available at:< http://www.saud epublica.ufc.br/imagens/u plo ads/dissertacoes/eb 17fe5a3c0 8d5a7a4fa95eb9d484c1a.pdf>. Accessed on: 20 July 2015.

SANTOS, C.A; JÚNIOR, F.O.L. Regional development and the formation of a polarised space in the southern mesoregion of Ceará: the municipalities of Crato, Juazeiro do Norte and Barbalha. IX Bahia Economics Meeting - SET. 2013.

SANTOS, M.; AREAS, M.A.; REYES, F.G. R. Pyrethroids-an overview. Alimentos e Nutrição Araraquara, v. 18, n. 3, p. 339-349, 2008. Available at <http://www.uff.br/toxicologiaclinica/Toxicologia%20dos%20Piretroides.pdf> Accessed on 04 September 2016.

SEBRAE NACIONAL. The vegetable market in Brazil. 18/11/2015. Available at: http://www.sebrae.com.br/sites/PortalSebrae/artigos/o-mercado-de-hortalicas-no-brasil,92e8634e2ca62410VgnVCM100000b272010aRCRD>. Accessed on 11 July 2016.

SILVA, S.L. et al. Survey of pesticide use in vegetables in Juazeiro do Norte-CE.VI SEPRONE, Campina Grande, 2011.

SIQUERIRA, S.L.; KRUSE, M.H.L. Agrotoxics and human health: the contribution of health professionals. Rev. esc. enferm. USP vol.42 no.3 São Paulo Sept. 2008.

SOARES. W. L. Use of pesticides and their impact on health and the environment: an integrated assessment of economics, public health, ecology and agriculture. 2010. 163 f. Thesis (Doctorate in Sciences in the area of Public Health and the Environment) - Oswaldo Cruz Foundation. Rio de Janeiro, 2010. Available at: <http://www.arca.fiocruz. br/bitstream/icict/2589/1/ ENSP_ Tese_ Soares_Wagne r_Lopes.pdf>. Accessed on: 23 July 2015

TERRA, F. H. B. The pesticide industry in Brazil. 2008. 171 f. Dissertation (Master's in Economic Development) - Federal University of Paraná. Curitiba, 2008. Available at: <http://www.economia.ufpr.br/Dissertacoes%20Me stra do/13 2%2 0% 20Fabio%20 Henrique%20Bittes%20Terra%20II.pdf>. Accessed on: 23 July 2015.

WAICHMAN, A. V. A Problemática do Uso de Agrotóxicos no Brasil: a necessidade de construção de uma visão compartilhada para todos atores sociais. Revista Brasileira de Saúde Ocupacional, São Paulo, v.37, n.125, 2012.

CHAPTER 8
ANNEXES

FORM TO OBTAIN INFORMATION ON

Researcher: ____________________________ -
Questionnaire number ______
Person Interviewed (Manager or owner) ______
Date: / /

Property location*: Georeferencing S 5 ' " W

_ . ..

* Crato

I - COMPANY IDENTIFICATION

Company name (or owner's name) ____________________
Address ____________________ Telephone: ____________
Property area ____________________ Cultivated area ____________
Type of crop** (m²) ____________________
** screened, greenhouse or open field
Staff currently employed: ____________ Gender: male ________ female _______
Year of implementation: ____________ Year of operation: ____________________

The company is: ( ) 1. independent ( ) 2. part of an association.
Which one? ____________________

Schooling of the interviewee:

Teaching
1.( ) Illiterate
2.( ) Incomplete primary education
3. ( ) Complete primary education
4.( ) High school incomplete
5.( ) Completed high school
6.( ) Incomplete university degree
7.( ) Complete university degree
8.( ) Foot-Graduation
Total

HORTICULTURE IN THE MUNICIPALITY OF CRATO

Workers' schooling:

Teaching	Number of staff employed
1.( ) Illiterate	
2.( ) Incomplete primary education	
3. ( ) Complete primary education	
4.( ) High school incomplete	
5.( ) Completed high school	
6.( ) Incomplete university degree	
7.( ) Complete university degree	
8.( ) Foot-Graduation	
Total	
Products	**Weekly production**
1.( ) Coriander	
2.( ) Chives	
3. ( ) Chilli	
4.( ) Chayote	
5.( ) Chilli pepper	
6.( ) Lettuce	
7.( ) Tomato	
8.( ) Sweet potato	

9.( ) Cassava	
10.( ) Other. Please specify.	

Products:

I - WORKERS' HEALTH, SAFETY AND WELL-BEING.

1. Have any health problems been reported by workers during crop management work?
   1. ( ) Yes 2. ( ) No If yes, which one(s)?________________________________

2. Is a person with first aid training always present during activities?
   1. ( ) Yes 2. ( ) No

3. Are there first-aid kits near the workplace?
   1. ( ) Yes 2. ( ) No

4. Do all workers use personal protective equipment (boots, hats, gloves) in their daily work?
   1. ( ) Yes 2. ( ) No

5. Is personal protective equipment (overalls, rubber gloves and boots, goggles and a mask) available for pesticide application?
   1. ( ) Yes 2. ( ) No

6. Do workers have access to basic services and facilities (toilets, drinking water) in the workplace?
   1. ( ) Yes 2. ( ) No

II - ENVIRONMENTAL MANAGEMENT

1. What is the source of water for use on the property?

( )1-	Deep well (semi artesian)
( )2.	Water from the water distribution company
( )3.	Dam water
( )4.	Water well
( )5.	Spring
( ) 6.	River. Please specify:__________

2. What kind of technology is used to irrigate the crops?

( )1.	Sprinkling
( )2.	Drip
( )3.	Hose
( ) 4.	Hydroponics

3. Does the company have a pesticide application registration system?
   IF YOU DO NOT USE PESTICIDES, LIST THE ALTERNATIVE PRODUCTS USED IN THE **GENERAL COMMENTS.**
   1. ( ) Yes 2. ( ) No

4. Have pesticide applicators received training to carry out their work?
   1. ( ) Yes 2. ( ) No

5. Have pesticide applicators ever read the package leaflet?
   1. ( ) Yes 2. ( ) No

6. Who recommends pesticide dosages?

( )1-	0 owner (reading the leaflet)

( )2.	0 seller
( )3.	A technician
( ) 4.	An agronomist

7. Do you reuse empty pesticide containers?
   1. ( ) Yes 2. ( ) No

   If you reuse, what for?______________________________________

8. What is the destination of empty pesticide containers? ________________________

9. Are pesticides stored in accordance with current legislation in Brazil (See NBR 9.843)?
   1. ( ) Yes 2. ( ) No

   Observations (specify non-conformities):

____

____

____

____

____

____

____

10. Types of pesticides used in your company.

Product	Purpose	Dosage	Culture

GENERAL COMMENTS:____

____

____

____

____

Product	Classification toxicological	Environmental classification	Suitable for use in horticulture	Class	Use in the company	Recommended dosage	Dosage used

MIX
Papier aus verantwortungsvollen Quellen
Paper from responsible sources
FSC® C105338

Printed by Books on Demand GmbH, Norderstedt / Germany